A PHOTOGRAPHIC HISTORY OF TURNER'S FIRST CENTURY

A PHOTOGRAPHIC HISTORY OF TURNER'S FIRST CENTURY

GREENWICH PUBLISHING GROUP, INC.

Produced and published by Greenwich Publishing Group, Inc.
Lyme, Connecticut

Design by Clare Cunningham Graphic Design

Library of Congress Control Number: 2002102175

ISBN: 0-944641-54-7

First Printing: March 2002

Photo on pg. 155, top, ©Tim Griffith. All other photographs courtesy of Turner Construction Company.

10 9 8 7 6 5 4 3 2 1

Building the Future

1902 – 2002

Turner Construction Company was founded on the simple belief that success is based upon the application of strong moral values, integrity, and the highest level of professionalism achievable. In 1902, these were the individual standards that were the fabric of a new company. They would prove to be the best assurance of Turner's longevity in a cyclic and unpredictable working environment. Through the years, Turner has been tested many times, and one can always conclude that these enduring core values, including being client-driven and people focused, are the reason the company still prospers after a hundred years in business. We are confident that by preserving these basic values, continuing to care for our employees, and remaining committed to community, we will have all the resources we need to build the future of the next century.

Preface

Turner Construction Company's Henry Chandlee Turner was born on a farm in Maryland in 1871, the second-eldest of a family comprised of six sons. He and his two youngest brothers, William W. and Joseph Archer (Archie), earned engineering degrees at Swarthmore College. Henry graduated in 1893 and later earned a graduate degree in civil engineering. He worked briefly in Philadelphia and then joined the U.S. Leather Company in New York City, working in the company's real estate department. Searching for his calling, Henry joined the Ransome Concrete Company in 1899. Ernest L. Ransome had been awarded patents for a reinforced concrete system based on using twisted steel bars in concrete to overcome concrete's lack of tensile strength. While this company was not successful, it provided valuable business knowledge and experience for Henry's future business endeavors.

Early in 1902, after raising $25,000, Henry decided to start his own company. It was a time when skilled labor was plentiful, the average workweek was fifty-nine hours and the average hourly wage was twenty-two cents. In the whole country, there were fewer than 150 miles of paved roads in the modern sense and only 8,000 registered automobiles. There were few buildings of any type that were more than ten stories in height, and few were designed in reinforced concrete. However, increasing land values dictated taller buildings and challenged engineers and architects to develop new methods of construction.

Henry C. Turner, *President and Founder*

Once he had decided to form his own company, Henry Turner successfully negotiated the rights to the Ransome process and the company began its first official job, constructing a cooper shop entirely of reinforced concrete for J. B. King & Company, a manufacturer. It was the first of many mutually beneficial projects built during the next few years, including a roof for a school, a bank vault, and several more structures for J. B. King. This early work generated a volume of $24,000 and earnings of $5,000 for the end of year results in 1902. Within two years, the company would see repeat work from three foundation clients that would assure the survival and prosperity of Turner's new company for years to come.

It is certainly worth noting that a large part of the early success of Turner Construction Company was directly attributable to a man who had spent many years with Ransome as an outside general superintendent, DeForest H. Dixon. Henry valued Dixon so highly that he was made a minority owner and an officer from the very beginning. It is highly apparent that Dixon was a brilliant field operations man; while Henry focused on the business model, selling new work, and setting the highest standards for the company, it was Dixon who managed labor, logistics, forming design, and fabrication of the twisted iron bars used to reinforce the pours. It is unlikely the company would have done as well as it did without this marvelous balance of management and distribution of responsibilities.

One of Henry's superintendents was William W. Turner, the second-youngest of Henry's five brothers. W. W., as he was called, had joined the company after graduation from Swarthmore College. He worked on a project for the U.S. Navy but resigned in 1907 to start a new and separate firm in Philadelphia, Turner Concrete Steel Company, which operated under the Ransome patents. He would ultimately employ his younger brother, Archie Turner, who graduated from Swarthmore College in 1905. Archie had worked for the Pennsylvania Rail Road in Altoona, Pennsylvania, and then the Ransome Company in Connecticut had employed him briefly before he joined his brother's new construction firm in Philadelphia as general superintendent. This was done after struggling with a difficult decision. At the time, both Henry and W. W. were seeking their youngest brother's talents. Ultimately, the catalyst and deciding factor resulted from not a business, but a personal preference. Archie's fiancée Helen Carré, also a Swarthmore graduate, was now living at her parent's home in Philadelphia. There were no hard feelings about that decision.

W. W.'s new company would also do quite well and thrived by building projects much the same as Turner Construction Company—until a major client defaulted on a contract with $200,000 still owed to Turner. W. W. chose to cancel the contract, as the idea of engaging in a legal battle was distasteful to him. Henry and W. W. agreed that after Turner Concrete Steel Company had finished its ongoing projects, it would cease operation and be absorbed into Turner Construction Company, a transition that was completed in 1919.

The story of Turner Construction Company is not just about a family-oriented business that two generations of Turners led and built; it is also about the marvelous and extremely talented young engineers and tradesmen that were attracted to a young and sure person with a vision. Henry Turner carried out that vision with and through the very dedicated efforts of those who believed passionately about the wonders of concrete and what it could do for a growing America. Henry was always aware of the reasons for his success and always supported and valued those passionate employees. Often, the actions adopted under Henry with the support of the Turner board were truly compassionate.

First- and second-generation Turners to serve in senior management included:

W. W. Turner, *Vice President and Member of the Board*

William W. Turner had for twelve years acted as founder and president of Turner Concrete Steel Company until it was absorbed by Turner Construction Company. He then became the general manager of the Philadelphia office. He retired during the early years of the Great Depression to make a place for his younger brother J. Archer Turner.

J. Archer Turner, *President and Chairman Elect*

J. Archer Turner was general manager of the Philadelphia office until 1941, when he became president, leading the company through the difficult and demanding years during World War II. Archie was elected chairman in 1946, but because of his untimely death, he never assumed the position.

H. Chandlee Turner, *President and Chairman*

H. Chandlee Turner, Henry Turner's oldest son, became president shortly after his uncle's death. Chan became president and chairman in 1954. He led the company to ever-greater levels of prosperity for more than twenty years.

Howard S. Turner, Ph.D., *Chairman & CEO*

Howard S. Turner was J. Archer's second eldest son. By 1967, Chan Turner had persuaded Howard to leave the outside corporate world to become the company's third president. He had never been directly associated with the firm. However, he had served as an outside board member since 1952. Howard became chairman and CEO upon Chan's retirement in 1969. Howard expanded the company, pushed it to become a public corporation, and saw the company reach a volume of $1 billion in revenues. Howard, who retired in 1978, was the last Turner family member to serve in the company's most senior ranks.

The Turners left a strong legacy built on hard work, the highest ethical standards, and a belief that the company's employees are assets to be treasured. The many significant leaders that have shaped and grown the company since 1978 have made sure that these values endure today. Today, Thomas C. Leppert, chairman and CEO of The Turner Corporation, and Robert E. Fee, president and CEO of Turner Construction Company, fully embrace the founding family's values and are wholly confident that this philosophy will enable the company to see its two hundredth birthday.

The photographs contained in this book are from the company's archive, a collection of leather-bound volumes (amazingly well preserved) that contain progress photos often used by the early leaders to represent work completed and also to illustrate methods of placing concrete. The practice of collecting and binding such photos lasted into the 1990s, when the volume of completed projects from multiple geographic office locations made the task too difficult to manage efficiently. Over the years, the books that were assembled, each one measuring three to six inches thick, grew in number to nearly 1,500, filling almost 270 cartons. The total number of slides and photos examined were in the 300,000 range, and the people who labored over the collection electronically scanned more than 20,000 images. These are now a permanent collection for use by Turner employees.

Much of this book is devoted to the formative years of Turner's first century and provides a look at the means, methods, workers, and staff employed in the early days of what would become a great company that played a substantial role in the building of America. It was a time of invention, great new technologies, and social change; and it was a time of the great wars and the Great Depression, which tested not only the mettle of the management of the company but also the perseverance of the whole country.

The intent of this book of photos is to share with the reader a look back at another time, when amazing feats of skilled labor and innovation created the manufacturing and industrial infrastructure so important to the growth of America. For many of those years, the workingman and management spent ten-hour days at their jobs, six days a week. There was no other expectation.

While the images are loosely organized by time period, the primary goal of the book is to present the best visuals from Turner's initial decades and to share unusual views of how the work was performed. In many ways, the projects were planned, organized, and conducted much the same as they are today. But as much as the process has remained the same, we innately know that life was harder, work was less safe, and compensation was much lower in the first part of the twentieth century than it is today. We can all be thankful that those things have changed. But more importantly, we can be thankful for the values, ethical standards, and focus on people that were so rigidly held to by Turner Construction Company—and that still survive today, after a century of excellence.

Medical Chambers Building, 140 East 54th Street, New York, New York, 1930

Henry Turner, pictured top center, never failed to reward dedicated, quality workmanship. In this picture, taken in 1931, the company recognizes the top performing workmen following completion of this project. *(1348)*

Note: The number in parentheses at the end of the caption denotes the official contract number.

THE 1900s

Opposite:
Load test on subway stairs, 135th and Lenox Avenue IRT Station, New York, New York, 1904

Henry Turner gambled on being able to build concrete staircases for less than what the city was paying for steel. He won, and the company went on to build more than fifty of them. *(52)*

Cooper's Shop, J. B. King & Company, Staten Island, New York, 1902

Built during Turner's first year, this was the first of more than twenty projects for J. B. King & Company. *(1)*

1904

Subway Stairs at the Astor Place IRT Station, New York, New York *(50)*

1904

Navy Laboratory in Brooklyn, New York *(65)*

1904

Subway stairs at the Brooklyn Bridge IRT Station, New York, New York *(75)*

Turner Construction Co.
CONCRETE-STEEL
CONSTRUCTION.
11 BROADWAY, NEW YORK.
OFFICIAL TEST
NEW YORK BUILDING DEPT.
TOTAL LOAD 96.000 LBS.
600 LBS. PER SQUARE FOOT

URNER CONSTRUCTION CO.
REINFORCED CONCRETE
11 BROADWAY
NEW YORK

Opposite:
Load and fire test for the Robert Gair Company warehouse and factory in Brooklyn, New York, 1905

At 180,000 square feet, the project for which this test was being performed was, in 1905, the largest concrete building in the United States. *(97)*

Warehouse and factory building No. 3 for the Robert Gair Company, Brooklyn, New York, 1905

These views of the Gair work show construction in progress, above, load testing, right, and finished building, left. *(98)*

More work for J. B. King & Company in Staten Island, New York, 1907

This photo shows formwork for the lower floors of a five-story building. *(129)*

Right:
J. B. King & Company Reservoir, 1909

This photo shows a view of 1909's predominant transportation mode. *(255)*

YMCA
NOV. 11, 18, 25, 3.30 P.M.
AUSPICES Y.M.C.A.

Eastman Kodak building, New York, New York, 1906

Here, Turner workmen are placing concrete for Eastman Kodak in New York. George Eastman tried without success to induce Henry Turner to build for him in Rochester, New York, too, where Eastman Kodak would develop its corporate headquarters. *(152)*

Turner for Concrete

Someone tells someone about "Turner for Concrete"

Warehouse
Great Atlantic and Pacific Tea Co.
Jersey City, N. J.

Timmis & Chapman, Architects

180 ft. x 121 ft.
9 stories and basement

This warehouse is one of ten buildings which Turner has constructed for The Great Atlantic & Pacific Tea Company—all on cost-plus-percentage contracts. As a test of a contractor's fair dealing, ability and reliability, percentage work forms a good criterion. 40% of our work is cost-plus-percentage contracts.

An even better basis of judgment is his record on repeat orders. Half of all our business is repeat orders—the other half is from new clients who come to us because of the good, quick, economical work we have done for others—"Some one tells some one".

191-22

Turner Construction Company
Buffalo 244 Madison Ave New York Boston

Great Atlantic and Pacific Tea Company warehouse, Jersey City, New Jersey, 1908

This is one of many, large warehouses built by Turner for the Great Atlantic and Pacific Tea Company (later A&P) in various parts of the country. The building itself is shown above, and its interior, including its fire-sprinkler system, is shown at right. *(191)*

Alsen Cement Company stockhouse, Alsen, New York, 1906

Fine grading and placing twisted-steel reinforcing bars are shown for the floor of Alsen Cement Company's stockhouse in upstate New York. *(246)*

1907

Power plant for the Great Atlantic and Pacific Tea Company warehouse, Jersey City, New Jersey, photo taken in 1908 *(199)*

1907

Warehouse for Frederick Loeser & Company Incorporated, Brooklyn, New York, photo taken in 1908 *(226)*

1910

The Promenade, stadium addition for Harvard University in Cambridge, Massachusetts

A concrete colonnade and 10,000 seats were added to Harvard's seven-year-old stadium in this earliest of the big sports jobs that Turner would do. *(261)*

THE 1910s

Checking the work, 1910

Turner was regularly investing resources for checking the performance characteristics of poured-in-place concrete. *(344)*

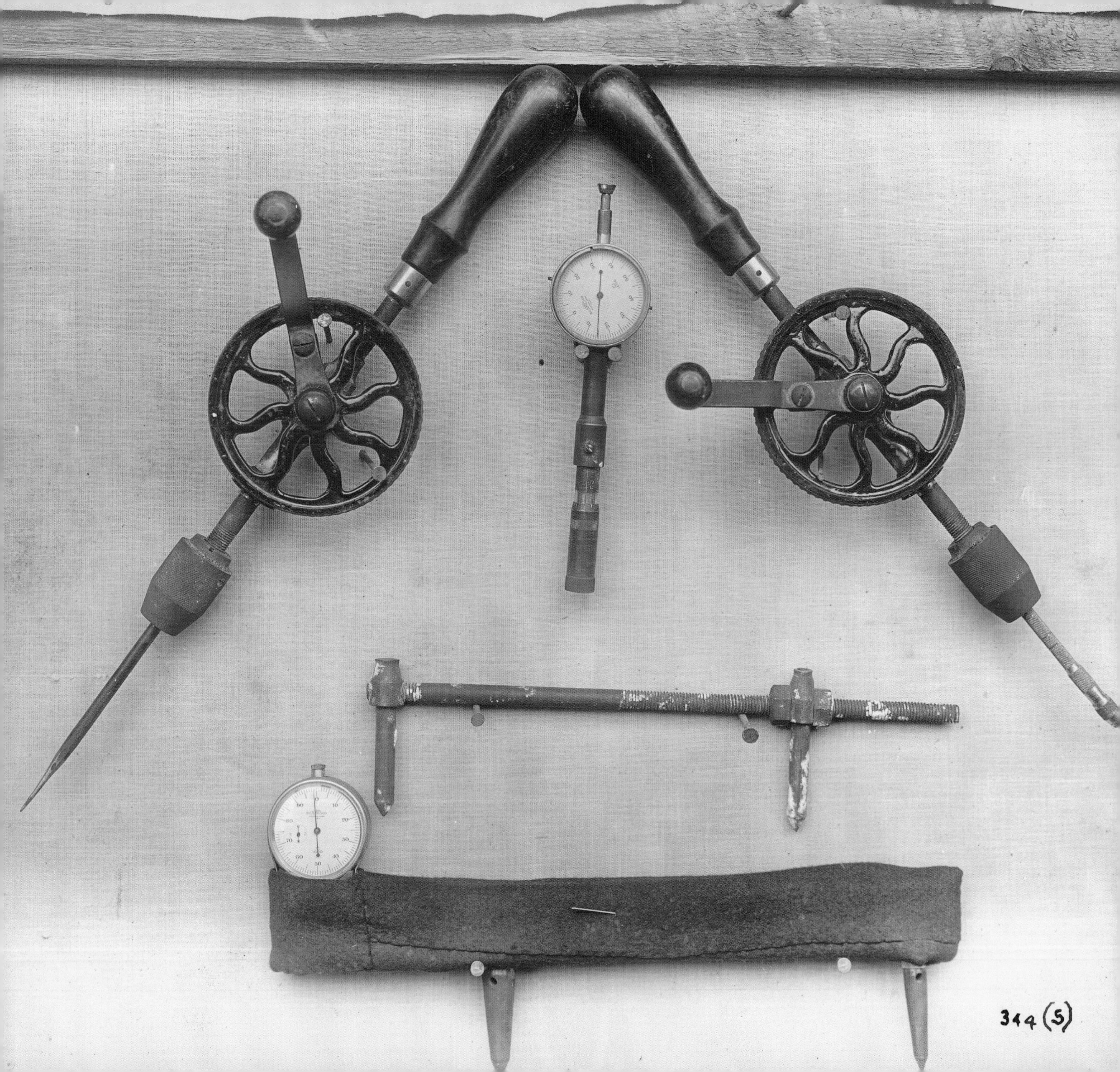
344 (5)

Opposite:
Tools for checking the work, 1910

Warehouse for the Baltimore and Ohio Railroad Company, New York, New York, 1913

Turner prepares for a complex pour in 1913. *(402)*

"TURNER *for* CONCRETE"

TURNER CONSTRUCTION COMPANY, *New York*

"Turner for Concrete" banner, 1902-1922

This was Turner's slogan until 1922, when it was discontinued to allow emphasis on finished construction. In the photograph above, a true steam shovel is shown.

Warehouse for Austin-Nichols & Company, Brooklyn, New York, 1912

When doing rock excavation, people had to be prepared to work! This photo of the site was taken in 1913. *(448)*

Service facility for the Texas Company (later Texaco), Long Island City, New York, 1915

Note the wood tower and cable system for transporting concrete across the water, a remote placement system not unlike concrete pumping systems in use almost a century later. *(497)*

Ransome concrete mixer, 1914

For nearly thirty years, virtually every Turner jobsite featured a patented Ransome barrel mixer operated by a steam engine (later by large, direct-current electric motors) together with a heavy-timber tower for elevating concrete that was chuted down to waiting, iron-wheeled Georgia buggies. *(453)*

1915

Factory of Seth Thomas Clock Company, Thomaston, Connecticut, photo taken in 1916 *(490)*

1915

Service facility for the Ford Motor Company, Buffalo, New York *(505)*

1916

Wax factory for the Vacuum Oil Company, Paulsboro, New Jersey *(574)*

Opposite and right: Merchant's Refrigeration Company warehouse, New York, New York, 1916

These two views from 1917 show foundation work at a busy, wet site. *(544)*

Belleville Warehouse Company, New Bedford, Massachusetts, 1916

The use of trucks on construction sites increased rapidly during and after World War I, but it would still be years before the last of the horse-drawn wagons would disappear. *(558)*

"TURNER FOR CONCRETE"
TURNER CONSTRUCTION CO.

NO
SMOKING

INVEST

CAMERA
CRAFT
N.Y.

687-C

"TURNER FOR CONCRETE"

A Turner Panorama

Navy and War Office Building, 1919

This panorama progress photograph illustrates the logistics required to build this project. Not only temporary office space for the government, but also tent quarters for imported skilled labor, a mess hall, a hospital, project field offices, storage, and fabrication facilities were all located strategically. Much of the material and equipment for each building was distributed by truck trestle. Note the base of the Washington Monument in the upper right corner. The Jefferson Memorial is located to the left of the site.

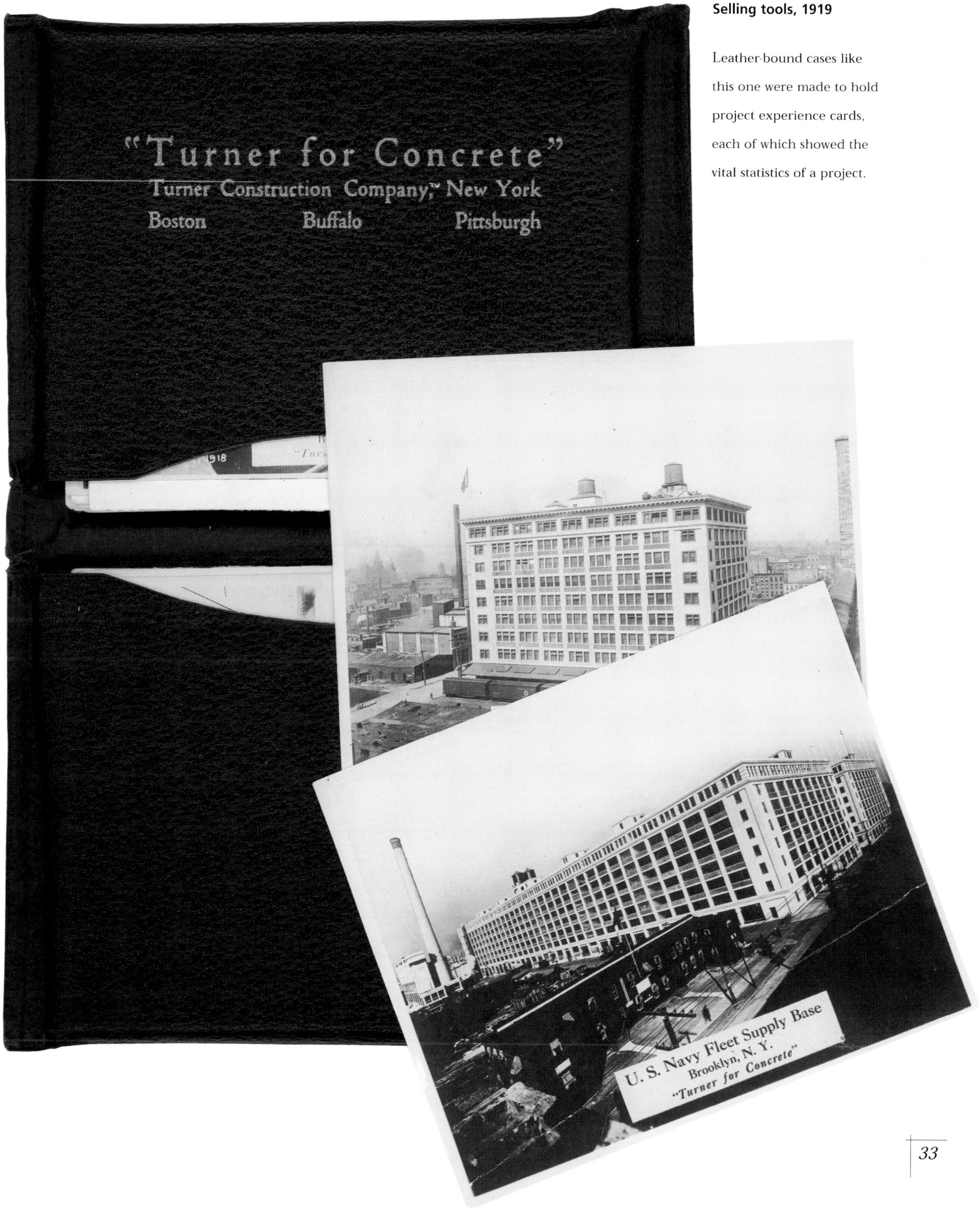

Selling tools, 1919

Leather-bound cases like this one were made to hold project experience cards, each of which showed the vital statistics of a project.

"Turner for Concrete"
Turner Construction Company, New York
Boston Buffalo Pittsburgh

U. S. Navy Fleet Supply Base
Brooklyn, N. Y.
"Turner for Concrete"

Page 32:
Staff on a Sanford Mills job, Sanford, Maine, 1919

From the beginning, Turner recruited its engineering staff from the best schools, including Cornell, MIT, Princeton, Swarthmore, and others. *(687)*

1917

Testing laboratory for the U.S. Bureau of Standards, Washington, D.C., photo taken in 1918 *(641)*

1918

Navy prison at Portsmouth, New Hampshire

This was Turner's first project for the corrections industry. *(652)*

1918

Navy and War Office Building, Washington, D.C., *(659)*

Navy and War Office Building, Washington, D.C., 1918

Turner's last World War I activity was dedicated by Franklin Delano Roosevelt when he was undersecretary of the navy. Completed in one year, most of the required skilled labor came from the northeastern states. The main entrance to the building is shown at left. *(659)*

"TURNER FOR CONCRETE"
684-A

Construction workers and foremen on a Sanford Mills job, Sanford, Maine, 1919

(687)

689-15
AUG. 26, 1919

Turner personnel on a job for Bristol-Myers, Hillside, New Jersey, 1919

This project added pharmaceutical-producer Bristol-Myers to Turner's clientele, which already included Squibb. Note the worker at center, in a humorous pose, strumming his saw like a banjo. *(689)*

Men working on Extension 6B to the Robert Gair Company factory complex, Brooklyn, New York, 1919

Placing reinforcing steel was commonly performed by Turner's crew. *(728)*

"TURNER FOR CONCRETE"
CAMERA CRAFT N.Y.
Y28-1
SEP. 5 19

Susquehanna Silk Mills, Milton, Pennsylvania, 1919

Turner often used photographs like these to illustrate steady progress. Here, the two pictures reveal dramatic changes between May and August of 1919. *(696)*

Opposite:
Ford Motor Company facilities in New York State, 1915, and in New Jersey, 1919

Turner's marketing materials during the 1920s often touted repeat clients by featuring finished project photos. *(505, 697)*

Albert Kahn, Architect FORD MOTOR CO., BUFFALO, N. Y. (Service Station) *Approximately 180,000 sq. ft.*

Albert Kahn, Architect FORD MOTOR CO., KEARNY, N. J. (Assembly Plant) *Approximately 765,400 sq. ft.*

Turner's Buffalo office, 1918

Here's a look inside Turner's first branch office, established in 1910 and closed in 1930.

Turner's Boston office, 1919

This photo illustrates the no-frills working environment that survived in the company offices until the late 1960s, when fashion and the influence of a corporate clientele encouraged a shift to more highly finished space.

RAYMOND
PILE
TURNER FOR CONCRETE
TURNER CONSTRUCTION CO.
Morgan
21152

The 1920s

International Textbook/Women's Institute Building, Scranton, Pennsylvania

The International Textbook Company specialized in publishing books for educating readers in the crafts. The American Women's Institute induced the company to produce a line of books targeted to female readers. A new building was needed, and Turner was engaged to build it. The picture at far left shows the crowd present at the groundbreaking, and the picture at left shows the building, completed in 1922. *(759)*

Eastern Pennsylvania Supply Company, Wilkes-Barre, Pennsylvania, 1921

Hand labor was rapidly being supplanted by machines, as this 1922 photo shows. *(819)*

Genesse Building, Buffalo, New York, 1922

This historically significant building, above, is a well-known landmark in full use today. The "Turner for Concrete" signs were beginning to give way to "Turner Construction Company, Builders" signs. Below is an issue of *Turner Constructor* magazine with the building featured on the cover. *(820)*

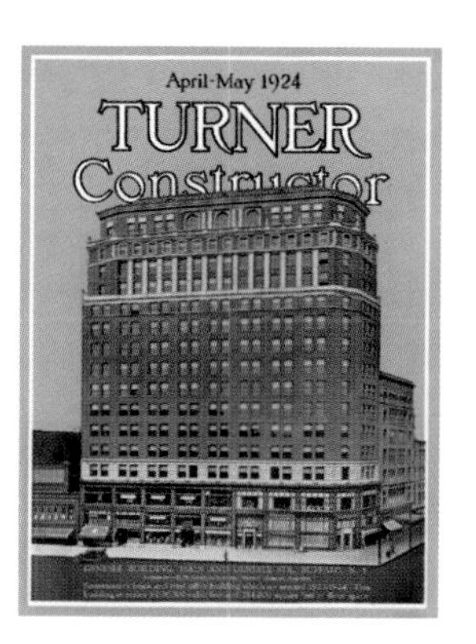

Construction of concrete docks for Havana Docks Company, Havana, Cuba, 1922

As trade with Cuba grew, the need for improved harbor facilities grew along with it. Turner built these municipal docks in a joint venture with its good friends, Raymond Concrete Pile Company, with whom Turner would do considerable other work as well. *(832)*

832-18 SEPT-19-1922
CAMERA CRAFT N.Y.

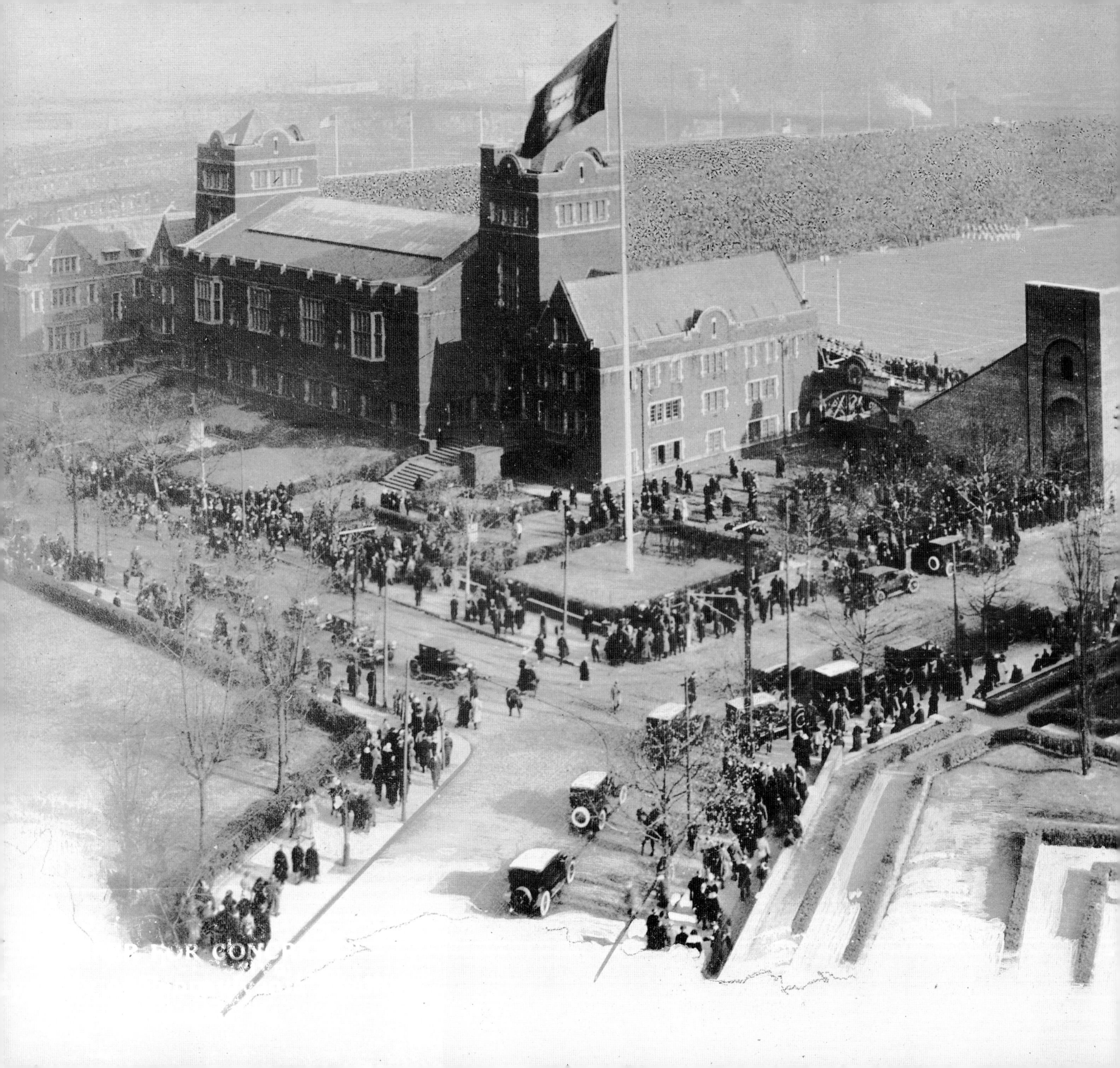

Franklin Field Stadium at the University of Pennsylvania, Philadelphia, Pennsylvania, 1922

Turner built Franklin Field under three separate contracts over several years. There were other large stadium projects during the same period and, toward the end of the century, a whole new generation of major sports complex construction. *(874)*

1922

San Juan hotel, Orlando, Florida *(860)*

1923

The Inn at Buck Hill Falls, Buck Hill Falls, Pennsylvania

Henry Turner had acquired a summer farm in Buck Hill Falls as a place for escaping the hot summers of Brooklyn and for golfing at a nearby course. Over time he built two houses on it, and one became a popular retreat in which Turner executives often combined working visits with golf and other recreation. Henry Turner was deeply committed to this Quaker resort community, and the company was the obvious choice to rebuild and expand the inn itself. This photo was taken in 1925. *(935)*

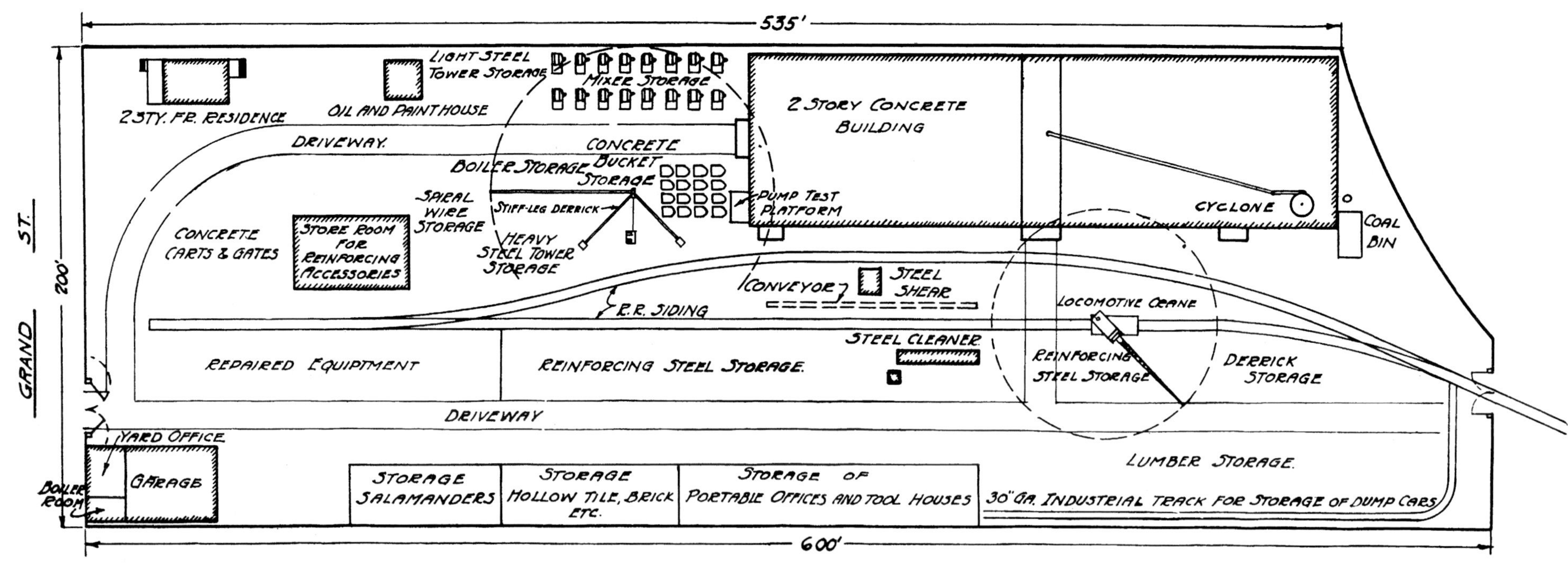

Plan of Yard

Turner yard and shop in Maspeth, New York, 1924

In general construction, equipment inventory is substantial and the need for space in which to store and maintain it is great. The Turner yard at Maspeth, shown here, covered 2 1/2 acres and included, in addition to a two-story, 38,000-square-foot shop building, a fully operative railroad siding, storage buildings, a residence, a garage, and a yard office. The shop building included a fully equipped wood mill, shown in the photograph, as well as well-equipped spaces for storing, maintaining, and repairing virtually every kind of construction equipment. *(963)*

August-September 1924 TURNER CONSTRUCTOR 11

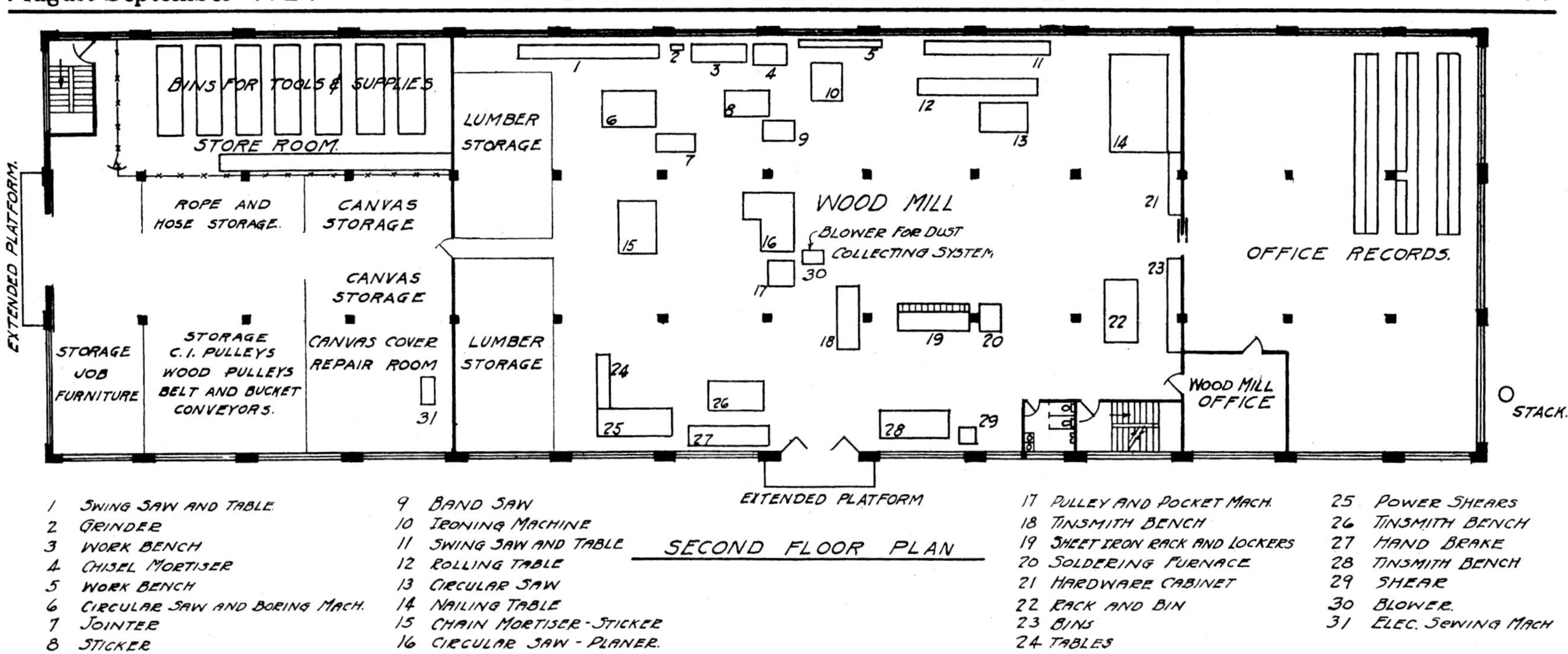

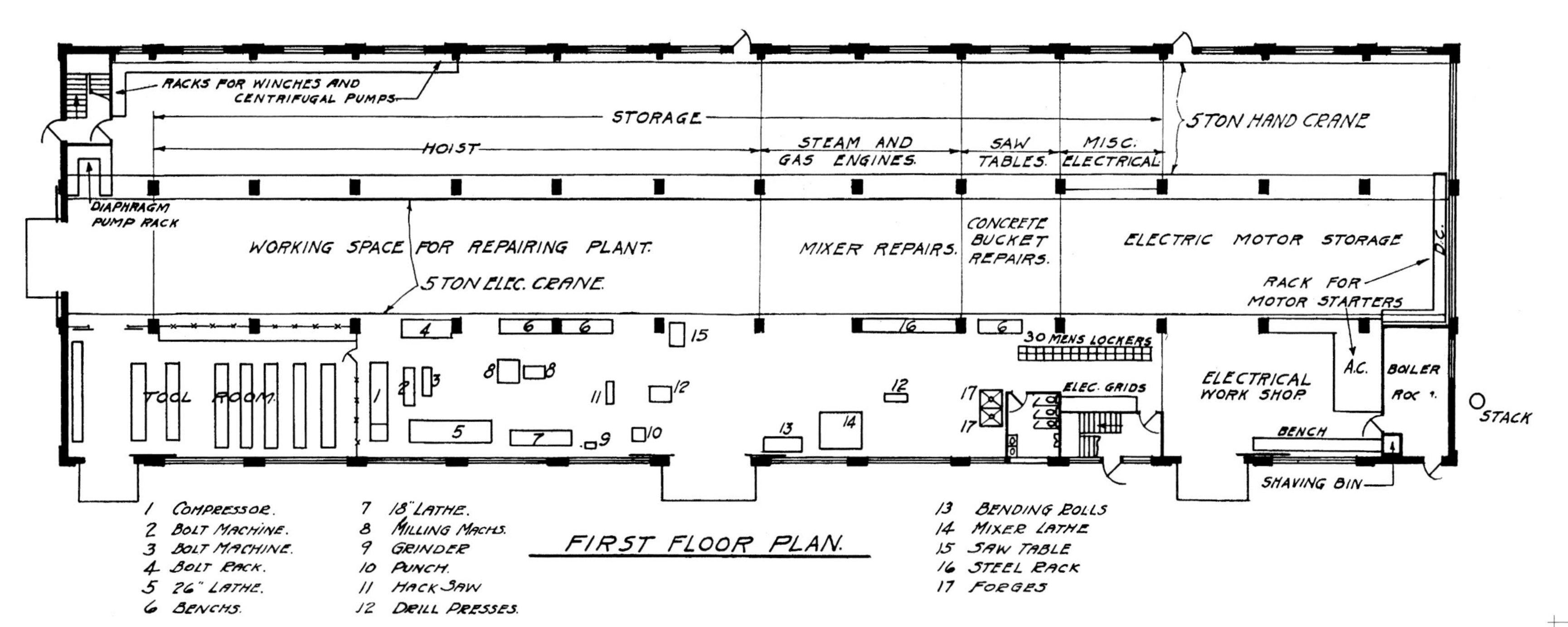

Opposite:
Koelle Greenwood Company, Service and Sales Building, Philadelphia, Pennsylvania, 1923

Among many projects for the industrial and manufacturing side of the automotive industry, Turner also built many showrooms, such as this Lincoln dealership, pictured in 1924. *(970)*

Equipment and crew at Turner's yard and shop, Maspeth, New York, 1924

In addition to its equipment, some of which is shown in the photograph above, the Maspeth yard and shop required a crew of close to fifty persons, shown right. *(963)*

LINCOLN
LINCOLN
CAMERA CRAFT N.Y.
970-1
JUNE-19-1924
Koelle Greenwood Company
Service and Sales Bldg.
Philadelphia, Pa.
Turner Construction Co.

Schoellkopf Field Stadium at Cornell University, Ithaca, New York, 1923

This photo of the completed stadium was taken in 1924.

(979)

Turner baseball team, 1925

...it wasn't all work!

Dennis Hotel, Atlantic City, New Jersey, 1924

By 1924 the new emphasis on finished work had been well established, and elegant hotels like the Dennis were the result. This photo of the finished interior was taken in 1928. *(1004)*

Opposite:
Venetian Hotel, Miami, Florida, 1925

Underneath all the elegance shown in this 1926 photo, there was still a concrete structure. *(1039)*

1039-19
March-19-1926
Venetian Hotel,
Miami Fla.
Turner Construction Co.

Massachusetts Mutual Life Insurance Company, Springfield, Massachusetts, 1925

Mass Mutual was in 1925 one in a long list of insurance companies for whom Turner would build. Innovation was continually at the heart of what Turner was doing. The photograph at left shows the elevator shaft in the Mass Mutual building, which was built to accommodate a truck for the distribution of materials on each floor. *(1064)*

The Breakers, Palm Beach, Florida, 1926

Completion of this elegant hotel confirmed that Henry Turner's decision to shift to finished construction had been fully implemented. These photographs, taken in 1927, give a good sense of the extraordinarily high level of finish and detail that characterized the project. *(1088)*

1926

Printing plant for the *New York Evening Journal*, New York, New York, photo taken in 1927 *(1095)*

1926

Home office of Provident Mutual Insurance Company, Philadelphia, Pennsylvania, photo taken in 1928 *(1101)*

1926

Buffalo–Fort Erie Bridge, Buffalo, New York *(1120)*

No. 1107,
July 8, 1927.

Railway Express Terminal for the Pennsylvania Railroad Company, Long Island City, New York, 1926

This view of the completed terminal dates from 1927.

(1107)

Factory building for Otis Elevator Company, Yonkers, New York, 1927

Here was a case where Turner benefited twice from changes in the society. As the population of the cities increased and the demand for high-rise construction grew, Turner built not only the new high-rise buildings, but also most of the factories in which Otis manufactured the elevators. *(1155)*

Cadillac showroom, New York, New York, 1927

Elegant cars like these were displayed in this Turner-built concrete building for almost forty years before the building was demolished (with great difficulty) to make room for Turner's Lincoln Center project. *(1157)*

New Jersey Bell Telephone Company headquarters, Newark, New Jersey, 1927

(1162)

James River Bridge, Newport News, Virginia, 1927

This major bridge job was unusual for Turner, which has tended to avoid heavy and highway work. But it was a profitable job involving the construction of almost six miles of bridges and five miles of connecting roads, forming a vital link in highway travel between Washington and Florida. The photos of the finished bridge, above and at left, were taken in 1928. *(1184)*

Lexington Hotel, New York, New York, 1928

Unlike Turner's other 1920s hotels, the Lexington, right, was a high-rise in the heart of New York's urban center. Above are some of the workers who distinguished themselves during the construction of the hotel, pictured in 1929, after it was completed, right. *(1215)*

Providence Terminal Market, Providence, Rhode Island, 1928

(1234)

United States Gypsum Company, East Chicago, Indiana, 1928

Turner kept moving dirt by combining the newest in equipment with the oldest.

(1228)

SEARS-ROEBUCK RETAIL STORE

ROCHESTER, N. Y.

NIMMONS, CARR & WRIGHT, ARCHITECTS

TURNER CONSTRUCTION CO, CONTRACTORS

4-15-1929

Sears Roebuck store, Rochester, New York, 1929

Spectators outnumbered workers on this job. *(1264)*

Dormitory for The Masters School, Dobbs Ferry, New York, 1929

This engine-powered "steam" shovel was a common sight in 1929, but its female operators weren't. *(1280)*

J.B. Lyon Company Printers, Albany, New York, 1929

Franklin D. Roosevelt, then governor of New York, was the guest of honor when this Turner job was dedicated in 1929. *(1290)*

Opposite:
***Wall Street Journal* Building, New York, New York, 1929**

Technicians measure the slump of fresh concrete. *(1298)*

1929

New York Central and Hudson River Railroad Company building, New York, New York *(1284)*

1929

Pratt & Whitney plant, East Hartford, Connecticut

Within another ten years, Turner would be building vast plants for this prolific builder of airplane engines. *(1296)*

1929

Ford Motor Company plant in Edgewater, New Jersey

This was Turner's seventh project for Henry Ford, who proceeded despite warnings that the economy was about to collapse. It would be many years before Ford would build another plant for the manufacture of automobiles. This photo was taken in 1930. *(1307)*

THE 1930s

Inland Terminal for The Port of New York Authority, 1931

The foundation contractor on this project subcontracted the concrete work to Turner at an extremely low price. But when the foundation work was finished, Turner was able to secure a profitable general contract for the superstructure, and a long and mutually beneficial relationship between Turner and the Port Authority followed. The detail at lower right shows the truck storage arrangement. *(1384)*

1930

Medical Chambers Building, Turner crew, 140 East Fifty-fourth Street, New York, New York, photo taken in 1931 *(1348)*

1931

Girard College Chapel, Philadelphia, Pennsylvania *(1377)*

1931

The Port of New York Authority, cargo terminal, New York, New York *(1384)*

Opposite:
U.S. Navy David Taylor Model Basin, Carderock, Maryland, 1937

This tunnel in the ship-testing basin was so long, and the tolerances for its construction were so exacting that the curvature of the earth had to be considered in its layout calculations. *(1514)*

Home office of New England Mutual Life Insurance Company, Boston, Massachusetts, 1938

(1530)

General Motors pavilion for the 1939 World's Fair, Flushing, New York, 1938

A complex structure in which the cars of the future were displayed, the General Motors *Futurama* exhibit was the fair's most popular. Fifteen years later Turner would build the General Motors exhibit at the 1964 World's Fair. *(1539)*

1939

Offices of Woodword & Lothrop, Washington, D.C. *(1535)*

1939

Storehouse for the U.S. Navy, Brooklyn, New York *(1596)*

1939

Pratt & Whitney Aircraft Company, wind tunnel and auxillary operations building, East Hartford, Connecticut *(1684)*

***Brantwyn*, residence of Pierre S. du Pont III, Rockland, Delaware, 1938**

(1555)

NX1384
Republic

Aircraft plant for Republic Aviation Corporation in Farmingdale, New York, and engine plant for Pratt & Whitney Aircraft Company, opposite, in Southington, Connecticut, 1939

By 1939 Turner was the country's most prolific builder of plants for airplane producers and for engine producers. The photo of the Pratt & Whitney plant at left was taken in 1942.
(1617 & 1673)

The 1940s

Above and opposite: Addition to the U.S. Navy David Taylor Model Basin at Carderock, Maryland, 1944

(1735)

Home office of the John Hancock Mutual Life Insurance Company, Boston, Massachusetts, 1945

The John Hancock Building was the first of several elegant buildings to be started in Boston just after the end of the war. This project, one of the first of such postwar contracts secured by Turner, marked the beginning of the company's rise to a position of dominance in the Boston marketplace. *(1773)*

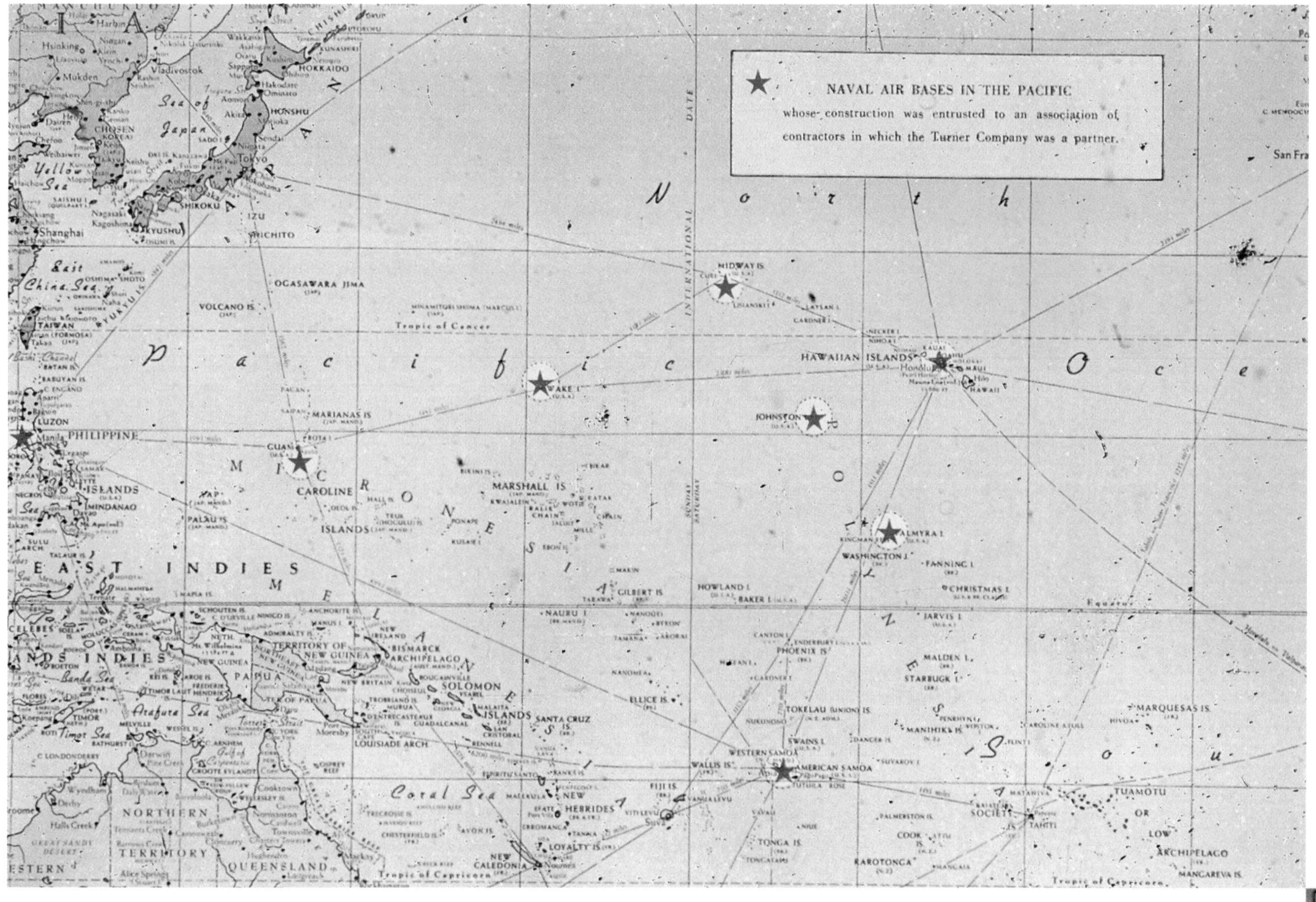

Pacific Naval Air Bases, 1940

Turner was the lead contractor in a joint venture organized to build a network of American air bases in the Pacific between 1939 and 1942, and the photograph opposite shows the work in progress in Hawaii. The map at left locates the work sites. The picture below left shows a building that was under construction in Pearl Harbor at the time of the Japanese attack. *(1572)*

Factory buildings for Carborundum Company in Niagara Falls, New York, left, and for American Can Company in Hillside, New Jersey, opposite, 1947

These were typical of Turner's postwar projects for firms that had been principal clients before the war. The photo opposite dates from 1948. *(1835 & 1841)*

TRANSIT MIXED CONCRETE
74
AMERICAN CAN COMPANY
HILLSIDE FACTORY BUILDINGS
TURNER CONSTRUCTION COMPANY
DATE T.C.CO.NO. 1841 VIEW NO.
6-11-48. 74

United Nations Secretariat Building, New York, New York, 1949

Turner was a member of a four part joint venture on this project, headed by the George A. Fuller Company and including Walsh Construction Company and Slattery Construction Company. *(1882)*

Boston University School of Theology, Boston, Massachusetts, 1947

This photo of the completed building was taken in 1949. *(1863)*

TURNER
CONSTRUCTION
COMPANY
PORT AUTHORITY
BUS TERMINAL
WORLD'S LARGEST BUS TERMINAL
TO OCCUPY THIS ENTIRE BLOCK FOR
CONVENIENCE OF COMMUTING AND
LONG DISTANCE BUS PASSENGERS.
READY BY THE END OF 1950
PORT AUTHORITY
BUS TERMI
W 40 ST
ONE-WAY

Bus terminal for The Port of New York Authority, New York, New York, 1949

Turner was awarded this contract when the owner questioned the ability of two lower bidders to complete the work satisfactorily and on schedule. Turner's product proved to be entirely satisfactory and was delivered on schedule. Both of these photos were taken in 1950. *(1916)*

THE 1950s

Chrysler Building East,
New York, New York, 1950

(1934)

**Cover of *Business Week*,
March 18, 1950**

Business Week featured a story announcing Chan Turner's selection to head the company, which he would continue to do for another twenty years.

Fairchild Camera and Instruments Corporation Building, Landia, New York, 1951

Here's an example of problem solving in the field.
(1974)

TURNER
CONSTRUCTION
COMPANY
THE NOISE YOU HEAR
IS THE SOUND OF PROGRESS!
MEMBER FEDERAL DEPOSIT INSURANCE CORPORATION
FARMERS BANK
of the
STATE OF DELAWARE
YOUR PATIENCE
AND UNDERSTANDING
MEAN EVERYTHING TO US
RMER'S BANK
DOVER, DEL.
DUPONT, ARCH.

New Jersey State Mental Hospital, Ancora, New Jersey, 1952

(1991)

Opposite:
Farmer's Bank of Delaware, Dover, Delaware, 1952

(1984)

AT&T Building, White Plains, New York, 1952

Designed to survive massive explosion or attack, the AT&T building and its enclosing walls were built entirely of reinforced, cast-in-place concrete. The project staff is shown above. *(2007)*

Connecticut General Life Insurance Building, Bloomfield, Connecticut, 1953

This high quality stone finish for the floor was a far cry from the trowelled concrete floors of the 1920s. *(2020)*

1952

The Evening Bulletin Building, Philadelphia, Pennsylvania *(1993)*

1953

Socony (later Mobil) Building on East Forty-second Street, New York, New York

Another project brought to Turner by the Galbreath interests, the Mobil Building became the site of Turner's own corporate headquarters when it was completed. *(2034)*

Cathedral of Mary Our Queen, Baltimore, Maryland, 1954

Built from 1954 to 1959, the Cathedral of Mary Our Queen is longer than St. Patrick's in New York, and its towers are sixteen stories tall. *(2059)*

Inland Steel Company Building, Chicago, Illinois, 1955

This was a prize project for Turner, which during the 1950s was seeking to establish itself as a major contender in Chicago. The upper photograph shows preparations for the topping-out ceremony in 1958, and the lower photograph shows the completed, nineteen-story building. *(2114)*

Chase Manhattan Bank Building, New York, New York, l956

(2119)

Alterations to the J.P. Morgan Building, New York, New York, 1955

(2124)

Addition to the *Daily News* Building, New York, New York, 1957

When work on the *Daily News* addition was started, Turner had just finished work on the Mobil Building, less than a block west of the *Daily News* site on Forty-second Street almost opposite Grand Central Station. By the end of the 1950s, Turner signs along Forty-second Street were familiar to every commuter. *(2169)*

Equitable Life Insurance world headquarters, New York, New York, 1957

Turner workmen finish up steel erection on the thirty-eighth floor of the new headquarters building for the 102-year old Equitable Life Assurance Society of the United States. With 1.6 million square feet of office area, the Equitable Building was said to be the largest privately owned office building in the world. *(2169)*

1956

American Airlines terminal, John F. Kennedy Airport, New York, New York *(2148)*

1958

Philharmonic Hall at Lincoln Center, New York, New York

To build Philharmonic Hall (later called Avery Fisher Hall), the joint venture of Turner, Fuller, Walsh and Slattery was re-established, this time with Turner in charge. This photo was taken in 1963. *(2209)*

1959

Expansion and reconstruction of LaGuardia Airport, New York, New York *(2230)*

633 Third Avenue, New York, New York, 1959

A Galbreath project, this building, pictured in 1961, later became the location of Turner's corporate headquarters. *(2218)*

Pan American Airways Passenger Terminal, John F. Kennedy Airport, New York, New York, 1958

Both the construction photo at right and the photo of the completed terminal below were taken in 1959. *(2190)*

Groundbreaking for Philharmonic Hall at Lincoln Center, New York, New York, 1958

Dignitaries abounded at the ceremonies and included (in this picture) President Dwight Eisenhower, Mayor Robert Wagner, and Commissioner of Parks Robert Moses. *(2209)*

The 1960s

Addition to the home office of New York Life Insurance Company, New York, New York, 1960

(2259)

Wisconsin-Clybourne Corporation Exchange International Bank, Milwaukee, Wisconsin, 1960

(2260)

Oglesby Towers, Chicago, Illinois, 1962

The snow didn't faze the groundbreaking crowd, it didn't faze the photographer, and, of course, it didn't faze Turner. *(2306)*

1961

The Gillette Company, Boston, Massachusetts *(2301)*

1962

General Motors pavilion at the 1964 World's Fair, Flushing, New York

This exhibit, also called *Futurama*, was completed just twenty-five years after Turner had built General Motors' pavilion at the 1939 World's Fair. This photo was taken in 1964. *(2313)*

1962

General Electric pavilion at the 1964 World's Fair, Flushing, New York, photo taken in 1964 *(2316)*

BETHLEHEM
BETHLEHEM
NEWSPAPERS
MAGAZINES
PENNSYLVANIA RAILROAD COMPANY.
REDEVELOPMENT PENNSYLVANIA STATION
NEW YORK, NEW YORK.
ARCHITECT. CHARLES LUCKMAN ASSOCIA
AGENT. TURNER CONSTRUCTION COMPAN
CONTRACT NO. 2321.

The new Madison Square Garden, New York, New York, 1963

Turner worked from 1963 to 1968, when these photos were taken, to complete the complex construction of this "most famous arena," below. Right, comedian Bob Hope wears a Turner hard hat. *(2383)*

Opposite:
Demolition of old Pennsylvania Station, New York, New York, 1962

Turner took the grand old station out of the way to make room for the modern Madison Square Garden and other new buildings. The demolition process, begun in 1962, was not completed until 1965, when this photo was taken. *(2321)*

Saratoga Race Track, Saratoga, New York, 1964

Here again John Galbreath had a role: he was a member of the New York State Racing Association. His confidence in Turner was such that it convinced his colleagues to engage Turner to manage the expansion and renovation of this facility. *(2400)*

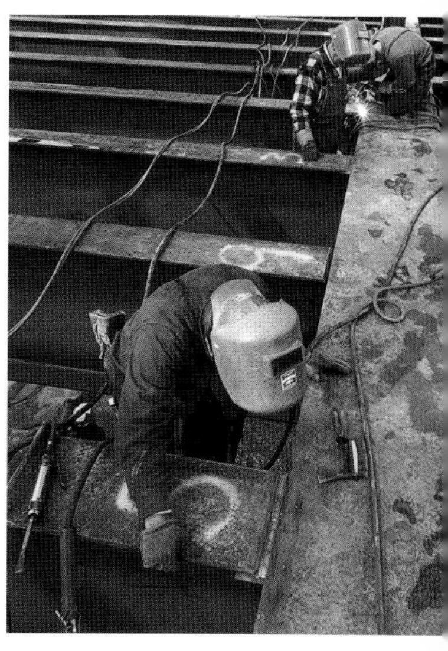

Above:
B. Altman & Company, Radnor, Pennsylvania, 1964

(2403)

Left:
Leisure and Creative Arts Center, George Williams College, Downers Grove, Pennsylvania, 1964

The sweeping timber arches and temporary scaffolding for the center's dramatic roof foreshadowed the modern art that would later be exhibited in the completed building. *(2415)*

Opposite:
Mei Foo Sun Chuen, Hong Kong, 1965

More than 13,000 housing units in ninety-nine separate high-rise buildings were built by Turner during the period from 1965 to 1978. It provided housing for about 75,000 persons. Mobil was the principal owner. *(2483)*

Renovation of the tower roof of the New York Life Insurance Company home office, New York, New York, 1965

The photograph above shows the renovation in progress. The photograph right shows the fully installed gold surface. *(2472)*

Plaza and surrounding buildings at Lincoln Center, New York, New York, 1965

When Philharmonic Hall was completed, the team that built it was disbanded and the remaining work at Lincoln Center was assigned to its individual contractors. Turner was awarded the lion's share of the work, and it included the New York State Theater, shown at right, and the Vivian Beaumont Theater, shown below, as well as the Library and the Museum. Picture at left shows the central plaza. *(2333 & 2334)*

1965

Robert A. Millikan Memorial Library at California Institute of Technology, Pasadena, California *(2485)*

1967

Carthusian Monastery, Arlington, Vermont, *(2561)*

1967

Shawmut Bank of Boston, NA, Boston, Massachusetts *(2600)*

By the 1960s, Turner was a major contender for the construction of new high-rise buildings in Boston.

New York Stock Exchange, New York, New York, 1966

The above photo shows the interior of 20 Broad Street as it appeared in 1969, after alterations were completed. *(2578)*

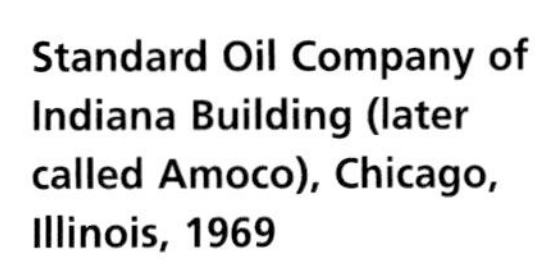

Standard Oil Company of Indiana Building (later called Amoco), Chicago, Illinois, 1969

A white marble tower when it took its place on the Chicago skyline, it is 100 feet shorter than the Empire State Building and is among the tallest buildings in the world. *(2719)*

1967

PepsiCo World Headquarters, Harrison, New York *(2606)*

Not every new world headquarters was a high-rise. In suburban Harrison, New York, the distinguished architect Edward Durell Stone designed this award-winning low-rise complex for PepsiCo and Turner built it.

1968

McGraw-Hill Building, New York, New York *(2674)*

1969

American Airlines terminal expansion, John F. Kennedy Airport, New York, New York *(2683)*

THE 1970s

Newspoint I, Long Island City, New York, 1970

Mayor John Lindsay, wearing a Turner hard hat, broke ground for a new printing plant for the *New York Daily News.*

(2754)

Dana World Headquarters, Ohio, 1970

Dana Corporation, a major auto parts manufacturer, was one of many projects built by the Lathrop Company for the automotive industry. Turner acquired Lathrop to expand geographically and enhance its market segment growth opportunities. Lathrop has been a remarkably strong contributor under the inspired leadership of its President Robert L. Maxwell. *(8600-L)*

1972

Sears Eastern Headquarters, Radnor, Pennsylvania *(2907)*

1973

Nabisco Global Headquarters, East Hanover, New Jersey *(2987)*

1973

Renovations to the Carnegie Institute, Pittsburgh, Pennsylvania *(3020)*

Opposite:
Sun Oil Company headquarters, St. Davids, Pennsylvania, 1975

(3107)

Ohio Bell Telephone Company, Cleveland, Ohio, 1977

(3339)

American Stock Exchange, New York, New York, 1977

Turner modernized the trading floor and completed other renovation work without interfering with trading operations. *(3344)*

John Fitzgerald Kennedy Memorial Library, Dorchester, Massachusetts, 1977

Turner built this elegant tribute to the thirty-fifth president of the United States. *(3377)*

**Analog Devices, Inc.,
Wilmington,
Massachusetts, 1977**

Dirty and difficult work, at left, probably gets a little easier, above. *(3390)*

1976

County of Ventura Pre-Trial Detention Center, Ventura, California *(3269)*

This was one of Turner's early projects for the corrections industry, a specialty in which the company would by the end of the century be the country's third most active builder.

1976

National Institute of Environmental Health Sciences, Research Building, Triangle Park, North Carolina *(3301)*

1977

NASA Wind Tunnel, Moffett Field, California *(3400)*

Three First National Plaza, Chicago, Illinois, 1978

A spectacular, 1.6-million-square-foot office building, this was one of many projects built by Turner for the Gerald Hines interests. *(3502)*

Bally Park Place Casino Hotel, Atlantic City, New Jersey, 1978

In this photograph, the old Marlborough-Blenheim Hotel is being demolished to make room for the new casino, but the Dennis, built by Turner during the 1920s, remained and was renovated. *(3523-A)*

WINZINGER
& CO
WINZINGER
WINZINGER
WINZINGER
TURNER CONSTRUCTION COMPANY
Bally Park Place Casino Hotel
Atlantic City, N.J.
Contract No. 3523A
Skidmore, Owings & Merrill, Architec
View Marlboro
Date: 20-9-78
No.
Photo by Hess Photography

THE 1980s

One Mellon Bank Center, Pittsburgh, Pennsylvania, 1980

This fifty-five-story office building was originally designed for the Dravo Corporation. *(3700)*

IBM building, New York, New York, 1980

Shown below is the IBM building under construction, which was also the site of Turner's patriotic display in support of American hostages in Iran. *(3500)*

Texas Commerce Tower, Houston, Texas, 1981

At seventy-five stories, Texas Commerce Tower was the tallest building west of the Mississippi. *(3531)*

Irving Trust Company, New York, New York, 1981

Turner's current president, Bob Fee, was superintendent on the construction of this building. *(3800)*

1980

General Electric Company, Schenectady, New York *(3691)*

This is yet another building Turner built for GE, for whom Turner has been doing major construction since 1912.

1980

Royal Terminal, New Riyadh International Airport, Riyadh, Saudi Arabia *(3725C)*

1981

Prudential Office Building, Louisville, Kentucky *(3806)*

United Airlines Passenger Terminal, O'Hare Airport, Chicago, Illinois, 1983

Considered to be state-of-the-art airport terminal design, the UAL terminal was said to be the largest building project in the country when it was under construction. *(4045)*

Office building at Fifty-third and Third, New York, New York, 1984

This building was called the "Lipstick Building" almost as soon as it was completed...and it's been known by that name ever since. *(4083)*

1982

Equitable Tower West, New York, New York *(3877)*

1983

Bank of America Technology Center, Concord, California *(3978)*

1983

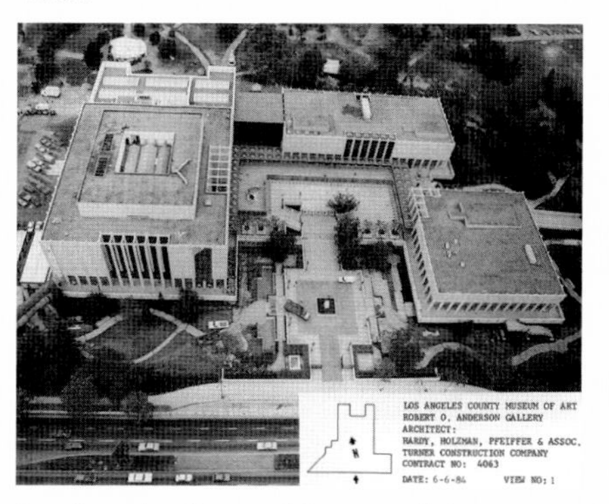

Los Angeles County Museum of Art, Los Angeles, California *(4063)*

Fisher Island Condominiums and Golf Course, Fisher Island, Florida, 1984

This project, top and above, has been growing since the 1980s, and the construction of new sections continues even today. *(4142)*

First Interstate World Center, Los Angeles, California, 1984

First Interstate, a seventy-five-story building designed to survive earthquake forces exceeding 8.3 on the Richter scale, contains almost 2 million square feet of interior space. *(4200)*

MORLEY
Turner
Construction
Company

Beckman Institute at the University of Illinois, Urbana, Illinois, 1986

(4370)

375 Hudson Street, New York, New York, 1987

This office building, owned by Tishman-Speyer, was the location of Turner's corporate headquarters until they relocated to Dallas, but it remains the site of Turner's New York office. The photograph shows Turner's reception space. *(4390)*

Terminal 2 at Los Angeles International Airport, Los Angeles, California, 1986

(4419)

THE 1990s

Cobo Conference/Exhibition Center, Detroit, Michigan, 1990

Turner's addition to the Cobo Center doubled the building's size, bringing it to about 2.4 million square feet and making it one of the largest convention centers in the country. *(4290)*

Amiri Diwan, Kuwait City, Kuwait, 1992

This 1.8-million-square-foot building was built for Kuwait's Crown Prince, Prime Minister, and Council of Ministers. The design and construction services were managed by a joint venture of Turner and Projacs, a Kuwait firm. *(57038)*

1992

Motorola, Chicago, Illinois, 1992 *(5046)*

1994

Genzyme, Allston, Massachusetts, 1994

Genzyme is a dynamic leader in biotechnology for whom Turner is doing additional work. *(5349)*

1995

American Hospital Dubai, Dubai, United Arab Emirates, 1995

This American-style hospital accomodates 100 beds. *(0113)*

Cincinnati Art Museum, Cincinnati, Ohio, 1992

Turner did the most extensive renovation in the 111-year history of the museum. *(5006)*

New England Medical Center, Boston, Massachusetts, 1991

Major additions and alterations on this project contributed to Turner's rising importance as a builder of health facilities. Turner was designated No. 1 in this specialty in *Engineering News-Record*'s national rankings later in the century. *(5179)*

Toledo Zoo-Great Ape Exhibit, Toledo, Ohio, 1992

(092001)

Gund Arena, Cleveland, Ohio, 1994

Home to the NBA's Cleveland Cavaliers, the Gund, built by the Lathrop company, would become the Midwest's pre-eminent location for basketball and large-scale public events.

(5231M)

Rock and Roll Hall of Fame, Cleveland, Ohio, 1995

This project gave Turner Construction Company a chance to prove that it was not limited to the well-worn paths of tradition. *(5421)*

Al Faisaliah Centre, Riyadh, Saudi Arabia, 1997

One of the tallest towers in the Middle East is the centerpiece of this 3.3 million square foot, mixed-use commercial, hotel and residential complex. The construction was managed by Turner. *(0168)*

Emirates Towers, Dubai, United Arab Emirates, 1996

These spectacular buildings were the tallest buildings in the United Arab Emirates upon completion. Turner served as the construction manager during the design and construction of these fifty-four- and fifty-six-story towers. *(0216)*

Museum of African American History, Detroit, Michigan, 1997

The Charles H. Wright Museum is one of the country's most important institutions devoted entirely to documenting, preserving, and educating the public in the history, life, and culture of African Americans. *(5444)*

New Jersey Performing Arts Center, Newark, New Jersey, 1997

(5453)

Opposite: Jurassic Park Theme Ride, Universal City, California, 1993

(5703)

Cumberland County Coliseum, Fayetteville, North Carolina, 1997

This complex, which includes a 13,500-seat sports arena as well as exhibition space, features entertainment fare ranging from the music of the Oak Ridge Boys to the "smash-mouth" football of the Cape Fear Wildcats. *(5671)*

1995

Avon High School, Avon, Ohio *(6002M)*

1995

Atlanta City Detention Center, Atlanta, Georgia *(5380)*

Design and construction were both Turner's responsibility on this 882-bed detention center, completed in twenty-four months.

1997

IRS Headquarters, Carollton, Maryland *(5549)*

1997

Ontario Convention Center, Ontario, California *(5628)*

1998

Aquarium of the Pacific, Long Beach, California, 1998 *(5742)*

Built by Turner in a joint venture with Kajima, this world-class marine exhibit displays marine life in forty-seven separate environments in a structure of more than 150,000 square feet. A complex facility designed and built to deal with exacting environmental demands, it filters almost a million gallons of water every hour.

1998

Douglas County Courthouse, Douglasville, Georgia *(6054)*

ETISALAT Telecommunications & Administration Building, Sharjah, United Arab Emirates, 1997

This is just one project in a multiple-building program of ten uniquely designed buildings. *(0138)*

International Terminal and Parking Facility at Atatürk Airport, Istanbul, Turkey, 1997

The prime minister of Turkey dedicated this approximately 350,000-square-foot international terminal, using this modernistic metal and glass structure to illustrate his view that Turkey will become one of the world's most prominent countries in the twenty-first century. Turner provided management services on this project. *(0185)*

Ritz Carlton, Doha, Qatar, 1999

A resort development that includes a 403-room five-star hotel, restaurants, and a marina-themed leisure island and water park. Turner managed the construction with joint venture partner Projacs of Kuwait. *(0207)*

THE 2000s

Seattle Cancer Care Alliance, Fred Hutchinson Cancer Research Center, Seattle, Washington, 2000

Turner has been building for this distinguished center of cancer research since 1996, and this laboratory/clinical care facility was its second project for the center. The Fred Hutchinson Center was founded in 1975 by William Hutchinson and named for his brother, an outstanding pitcher for the Detroit Tigers baseball team who died of cancer in 1969. *(6596)*

Northeastern University, student housing and parking, Boston, Massachusetts, 2000

(6720)

BOX Office

Anaheim Convention Center expansion, Anaheim, California, 2000

Turner completed an approximately 200,000-square-foot expansion to the largest convention center on the West Coast. *(6234M)*

Paul Brown Stadium, Cincinnati, Ohio, 2000

This Turner-built stadium is the home of the Cincinnati Bengals football team. *(6258)*

Biogen 8 Development Building, Cambridge, Massachusetts, 2000

Turner built this laboratory and research facility for Biogen, which was founded in 1978 and is the world's oldest genetic research firm. *(6739)*

Renovations to Ohio State University's Stadium, Columbus, Ohio, 1999-2001.

Under Turner's management, a monumental feat of renovation was accomplished that included lowering the playing field itself by more than fourteen feet .The staged project generated 11,000 additional seats and added new boxes, a new scoreboard, new concession stands, and the like. *(6594C)*

Kansas Speedway, Kansas City, Kansas, 2000

(6329)

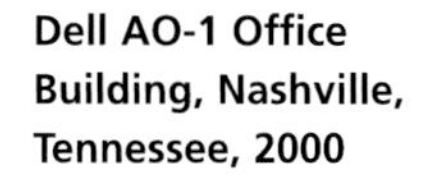

Dell AO-1 Office Building, Nashville, Tennessee, 2000

This project received the Associated General Contractors of America "Build America" award in 2001. Turner's subsidiary Turner Universal of Nashville was the builder. *(3027)*

OHIO STATE
BUCKEYES

Arctic Ring of Life Exhibit, the Detroit Zoo, Royal Oak, Michigan, 2001

The four-acre Arctic Ring of Life features outdoor and indoor exhibits that include a clear underwater tunnel that allows guests to walk through the polar bear pool while the bears swim above. This is billed as the largest exhibit of its type in the world.

Bear Stearns World Headquarters, New York, New York, 2001

One of the most spectacular buildings in the array of major office towers built by Turner during the recent renaissance of commercial construction in New York City. *(6411A)*

Invesco Field at Mile High, Denver, Colorado, 2001

The new home of the Denver Broncos football team. Turner has recently established a pattern of completing one of these mammoth sports stadiums each year. *(6677D)*

In the next hundred years, Turner will remain the industry leader, the builder of choice. Turner will achieve its goals in remaining faithful to the principles set forth by Henry C. Turner in 1902. And as in the past and present, Turner people will be the reason the company continues to grow and prosper.